Report NO. SPO-2009-005   July 24, 2009

Inspector General
United States Department of Defense

Assessment of Electrical Safety In Afghanistan

General Information

Forward questions or comments concerning this assessment report and other activities conducted by the Office of Special Plans & Operations to:

Office of the Assistant Inspector General
for Special Plans & Operations
Department of Defense Office of Inspector General
400 Army Navy Drive
Arlington, VA 2202-4704
or
E-mail: spo@dodig.mil

An overview of the Office of Special Plans & Operations mission and organization and a list of past evaluations and future topics is available at http://www.dodig.mil/spo.

Additional Information and Copies

To obtain additional copies of this report, visit the Web site of the Department of Defense Inspector General at http://www.dodig.mil/spo/reports or contact the SPO Secondary Reports Distribution Unit at (703) 604-8833 (DSN 664-8833) or fax (703) 604-9769.

If you suspect Fraud, Waste, Abuse, or Mismanagement in the Department of Defense, please contact:

DEPARTMENT OF DEFENSE

To report fraud, waste, mismanagement, and abuse of authority.

Send written complaints to: Defense Hotline, The Pentagon, Washington, DC 20301-1900
Phone: 800.424.9098   e-mail: hotline@dodig.mil   www.dodig.mil/hotline

INSPECTOR GENERAL
DEPARTMENT OF DEFENSE
400 ARMY NAVY DRIVE
ARLINGTON, VIRGINIA 22202-4704

July 24, 2009

MEMORANDUM FOR U.S. CENTRAL COMMAND
U.S. FORCES-AFGHANISTAN
DEFENSE CONTRACT MANAGEMENT AGENCY

SUBJECT: Assessment of Electrical Safety in Afghanistan
(Report No. SPO-2009-005)

We are providing this report for your information and use. We performed the assessment as a self initiated review. We considered management comments on our preliminary observations and a draft of the report in preparing the final report.

Comments on the preliminary observations and the draft of this report conformed to the requirements of DoD Directive 7650.3 and left no unresolved issues. Therefore we do not require any additional comments.

We appreciate the courtesies extended to the staff. Please direct questions to Brett Mansfield, Project Manager, at (703) 604-8302 (DSN 664-8302). If you desire, we will provide a formal briefing on the results.

Kenneth P. Moorefield
Assistant Inspector General
Special Plans & Operations

Report No. SPO-2009-005 (Project No. D2009-D00SPO-0192.000)　　　July 24, 2009

Results in Brief: Assessment of Electrical Safety in Afghanistan

What We Did
We assessed the effectiveness of command efforts to ensure the electrical safety of Department of Defense occupied and constructed facilities in Afghanistan. We visited a series of sites throughout Afghanistan, reviewed current efforts to assess electrical safety, and performed electrical assessments.

What We Found
U.S. Central Command (CENTCOM), U.S. Forces-Afghanistan (USFOR-A), and Combined Joint Task Force (CJTF)-101, are all aware of the risks associated with the electrical infrastructure within Afghanistan and have taken steps to address these issues. Despite these positive steps forward, additional coordinated efforts need to be undertaken to ensure electrical safety. Specifically, we found:
- Potentially dangerous situations that required immediate attention at Camp Brown and Forward Operating Base (FOB) Spin Boldak.
- The need for a full-time cadre of individuals dedicated to electrical safety throughout Afghanistan, a comprehensive inventory of U.S. controlled facilities in Afghanistan, and a comprehensive inspection plan.
- A need for an organization with authority to grant waivers to the National Electrical Code.
- A lack of education for service members regarding electrical safety, incident reporting, and personal responsibility.
- A need for additional Contracting Officer's Representatives (COR) and other oversight personnel to oversee electrical work being performed in Afghanistan.

What We Recommended
- USFOR-A take immediate action to correct the electrical deficiencies at Camp Brown and FOB Spin Boldak.
- USFOR-A dedicate a full-time cadre of personnel to electrical safety, develop an inventory of U.S. controlled facilities, and develop and execute a comprehensive electrical inspection plan for U.S. occupied facilities in Afghanistan.
- USFOR-A appoint an "Authority Having Jurisdiction" to grant waivers to the National Electrical Code.
- USFOR-A include training on electrical safety, incident reporting, and personal responsibility as part of pre-deployment and in-theater training.
- USFOR-A and the Defense Contract Management Agency (DCMA) identify and train the individuals needed to meet the COR and oversight personnel requirements.

Management Comments
CENTCOM provided comments to our preliminary observations and recommendations and to a draft of this report, which included planned and ongoing actions by USFOR-A, on May 11, 2009, and June 29, 2009. DCMA provided comments to a draft of this report on July 9, 2009.

Those comments have been incorporated into this report. Management generally concurred with all of our observations and recommendations. We consider management comments to be responsive to the recommendations and no additional comments are required. Please see the recommendations table on the back of this page.

Report No. SPO-2009-005 (Project No. D2009-D00SPO-0192.000) July 24, 2009

Recommendations Table

Management	Recommendations Requiring Comment	No Additional Comments Required
U.S. Central Command		1, 2, 3, 4, 5, 6, 7, 8, 9, 10, 11, 12
U.S. Forces Afghanistan		1, 2, 3, 4, 5, 6, 7, 8, 9, 10, 11, 12
Defense Contract Management Agency		11

Table of Contents

Results in Brief	i
Introduction	1
Objectives	1
Methodology	1
Background	1
Positive Actions Regarding Electrical Safety	2
Observations and Recommendations	
1. Electrical Issues at Camp Brown and FOB Spin Boldak	5
2. Comprehensive Inventory of U.S. Controlled Facilities in Afghanistan	8
3. Comprehensive Electrical Inspection Plan for U.S. Controlled Facilities in Afghanistan	10
4. Resources to Inspect, Detect, and Correct Electrical Deficiencies	12
5. Level of Recordable Electrical Accidents in Afghanistan	14
6. Full-Time Cadre Dedicated to Electrical Safety in U.S. Controlled Facilities in Afghanistan	15
7. Cap on Use of Operations and Maintenance Funds for Minor Construction in Afghanistan	16
8. Authority Having Jurisdiction to Grant Waivers to the National Electrical Code in Afghanistan	18
9. Training Soldiers on Electrical Hazards and the Reporting Process	20
10. Use of Unlisted Electrical Components	21
11. Qualified Contracting Officer's Representatives for Review of Electrical Work	22
12. Re-wiring of New Ablution Units at Kandahar Air Field	24
13. Kandahar Air Field Power Plant	26
Appendices	
A. Scope and Methodology	27
B. Additional Photographs	31
C. Legislative Proposal Concerning Minor Military Construction Projects	37
D. Management Comments to Preliminary Observations	43
E. Management Comments to Draft Report	49
F. Report Distribution	57

Introduction

Objectives

On March 31, 2009, we announced the Assessment of Electrical Safety in Afghanistan. The objective of this assessment was to review the effectiveness of command efforts to ensure the electrical safety of Department of Defense occupied and constructed facilities in Afghanistan.

Methodology

We examined mainly qualitative data during this project. The qualitative data reviewed consisted of individual interviews, direct observations, and documents.

On April 19, 2009, we began a one week assessment of the electrical safety of DoD facilities occupied or constructed by U.S. personnel and contractors in Afghanistan. The assessment team performed work at Bagram Airfield (BAF), Forward Operating Base (FOB) Altimur, FOB Sharana, Kandahar Airfield (KAF), Camp Brown, FOB Tarin Kowt, FOB Spin Boldak, Camp Phoenix, and Camp Eggers.

We used a subject matter expert (senior electrician from A Co., 249th Engineer Battalion (Prime Power)) to assess the electrical safety and code compliance of facilities visited at various locations. The team's subject matter expert performed limited assessments of electrical components at select facilities in Afghanistan. The subject matter expert did not conduct full electrical inspections of each facility due to time constraints in country.

Subsequent to departing Afghanistan, the team provided U.S. Central Command (CENTCOM) briefing charts with the team's preliminary observations and recommendations. We requested CENTCOM provide comments regarding our observations and recommendations. The Command provided preliminary comments on May 11, 2009. Those comments were incorporated into the draft and final reports. See Appendix D for comments provided by CENTCOM to the preliminary observations and recommendations. CENTCOM and the Defense Contract management Agency (DCMA) provided comments to the draft report on June 29, 2009, and July 9, 2009, respectively. See Appendix E for comments to the draft report provided by CENTCOM and DCMA.

Background

The Assessment of Electrical Safety in Afghanistan is a self-initiated review by the DoD Office of Inspector General (OIG). The 101st Airborne Division, Regional Command East (Combined Joint Task Force [CJTF]-101), Bagram, Afghanistan, stated that they established Task Force Protecting Our Warfighters and Electrical Resources (TF POWER) to evaluate and plan for safe power in Afghanistan.

TF POWER was established to "prevent the loss of life and government property through immediate and long-term measures that will significantly reduce the number of electrical and fire incidents throughout the combined/joint operations area."

These activities include:

- assessment of existing and new electrical resources and correction of deficiencies found
- inspection/re-inspection of electrical resources throughout the Area of Operations
- education on electrical and safety issues
- awareness-building and information dissemination through the Web and other media
- ensuring availability of National Electrical Code-compliant materials
- data collection, storage, and analysis

TF POWER used contractors to review and identify electrical deficiencies to include life, health, and safety issues at FOBs. According to TF POWER representatives, as of April 18, 2009, TF POWER tracked electrical inspections at 216 of 257 FOBs and completed 100 percent inspections of 16 bases. Additionally, TF POWER utilized standardized checklists for performing electrical reviews.

Positive Actions Regarding Electrical Safety

During our review, we noted awareness of the issues and positive actions taken by the command in theater to increase electrical safety in Afghanistan. Those actions included establishing a theater-wide electric code, ongoing construction and remediation, the establishment of TF POWER, DCMA resource assessment, and local Inspector General inspections.

Electrical Awareness
The Command produced a series of electrical safety bulletins, an inspection guide for safety officers, established the National Electrical Code (NEC)[1] as the electrical code within theater (by Fragmentary Order), and written guidance to restrict the sale of unlisted power strips in the Combined/Joint Task Force Operational Area.

New Construction and Corrective Work
The assessment team's subject matter expert reported that the electrical work for new construction that we observed was NEC compliant. Because the work observed was not complete, a full assessment was not possible. All plans and blueprints viewed, as well as the workmanship, were in accordance with NEC 2008. Additionally, we noted a number of instances where corrective actions had been taken to remediate previously dangerous or non-compliant situations. The following is a series of photos representative of the new construction and remediation we observed.

---

[1] The NEC is a set of standards published by the National Fire Protection Association (NFPA) for the safe installation of electrical wiring and equipment. The NEC is approved by the American National Standards Institute (ANSI) as ANSI NFPA 70. While the NEC is not itself a U.S. law, its use is commonly mandated by state and local law.

Figure 1. Panel Box installed by a Seabee in a B-hut, which was still under construction at Kandahar Airfield.

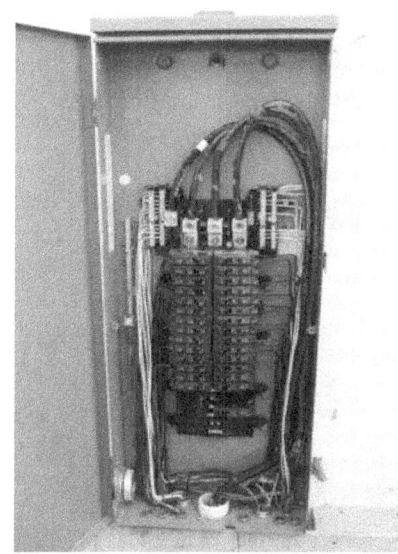

Figure 2. Color coding of wires by a contractor with tape to bring box into NEC compliance at FOB Altimur.

Figure 3. Temporary fix performed by contractor to protect personnel from coming into contact with exposed wires at Camp Eggers.

Figure 4. Local contractor work performed on the newly constructed Joint Operations Center at Camp Phoenix.

DCMA Actions
DCMA performed a workload assessment to determine their current staffing needs. Based on the results of that assessment, DCMA requested 45, and received approval for an additional 43 personnel, at least 30 of which had already been identified by name at the time of our site visit.

IG Command Inspections
CJTF-101 and 143rd Expeditionary Sustainment Command IGs incorporated an electrical component into their normal base inspections. The main focus is on life, health, and safety issues. Although the IGs are not trained electricians, they focus on apparent deficiencies. Using this model, CJTF-101 IG identified and reported a number of issues.

Observation 1. Electrical Issues at Camp Brown and FOB Spin Boldak

We observed electrical issues at Camp Brown and FOB Spin Boldak involving grounding, bonding, circuit protection, and personnel protection. The majority of the wiring and panels were not properly grounded or bonded, if at all; distribution panel doors were missing; electrical components were unprotected; and improper electrical insulators were used.

The following pictures represent observations made at Camp Brown and FOB Spin Boldak by the DoDIG assessment team, which was augmented by an electrical subject matter expert. The captions represent opinions by that subject matter expert. Appendix B shows additional photographs of electrical work in Afghanistan.

Figure 5. Camp Brown: Unprotected electrical component is an electrocution hazard.

Figure 6. Camp Brown: Electrical re-wiring, to correct work performed by an untrained worker, was performed by a Seabee to prevent injury or property damage.

Figure 7. Camp Brown: The undersized wire will melt before the breaker trips causing a fire and blast hazard. Panel on the left has no dead front causing a blast hazard.

Figure 8. Spin Boldak: A lack of grounding and proper bonding and undersized wires create a shock hazard.

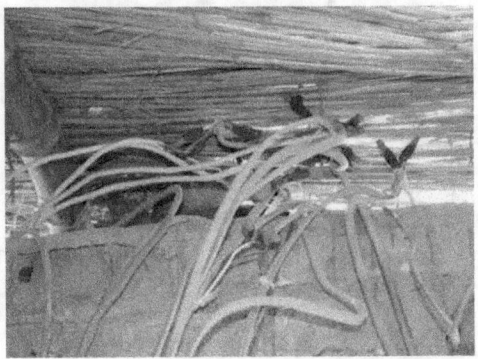

Figure 9. Spin Boldak: Exposed splices directly below an exposed straw roof create a fire hazard, and the splices are not weather-protected.

Figure 10. Spin Boldak: Connecting wires are undersized making them a fusible link, and there is no wire protection around the knock-outs at the bottom of the panel.

Figure 11. Spin Boldak: Panel doors or dead fronts missing from both panels, which increase potential for direct contact by personnel to energize circuits, which could result in arc blasts, shocks, and property damage.

Figure 12. Spin Boldak: Taped splices (unlisted connectors) are exposed to the weather and cause an increased risk of electrocution during wet weather.

Figure 13. Spin Boldak: A sock used as an electrical insulator and an exposed splice is a fire hazard.

The observed conditions required immediate correction or would likely result in significant safety issues. According to the assessment team's subject matter expert, the lack of (or improper) grounding or bonding causes a shock hazard to personnel. In the absence of distribution panel doors or dead fronts for personnel protection, there was potential for direct contact by personnel to energize circuits, which could result in arc blasts and property damage. Further, unprotected electrical components may result in electrocution hazards and the use of improper electrical insulators cause fire hazards.

Recommendation, Client Comments, and Our Response

1. We recommended that USFOR-A take immediate action to correct the electrical deficiencies at both locations.

Client Comments to Observation and Recommendation

CENTCOM concurred. CENTCOM further stated that as of May 11, 2009, USFOR-A corrected the Life Safety deficiencies. Additionally, CENTCOM stated that the estimated completion dates for work related to major discrepancies at Camp Brown and all electrical work at Spin Boldak was May 30, 2009. On June 29, 2009, CENTCOM stated that all Life Safety issues were completed at Spin Boldak and the completion of remaining electrical deficiencies at Spin Boldak was scheduled for September 16, 2009.

Our Response

CENTCOM comments were responsive. We request updates upon completion of the work at Camp Brown and FOB Spin Boldak.

Observation 2. Comprehensive Inventory of U.S. Controlled Facilities in Afghanistan

We observed a lack of a comprehensive inventory of U.S. controlled facilities in Afghanistan. Based on meetings with personnel from USFOR-A and CJTF-101, there was no comprehensive inventory of U.S. controlled facilities in Afghanistan. At the time of our visit, facilities with Operations & Maintenance (O&M) performed under the Logistics Civil Augmentation Program (LOGCAP) contracts were tracked by contractors on a density list, which detailed each facility by type and level of maintenance required. Facilities not covered under LOGCAP were not included on the density lists, leaving gaps in the Department's ability to assess, plan, and track electrical repairs within Afghanistan.

According to CJTF-101, at the time of our visit there were a total of 257 FOBs in Afghanistan. CJTF-101 personnel provided summary information indicating at least 12,246 facilities under LOGCAP contracts in Afghanistan. Additionally, not all facilities on a FOB covered by LOGCAP were contained in the contractor listing. If the contractor was not assigned O&M responsibility on a facility, it would not be included in the listing for that FOB.

Table 1. Current Number of Facilities Under LOGCAP

CONTRACT	Number of Bases	Number of Facilities
LOGCAP III	82	11,361
LOGCAP IV	12	885
Total		**12,246**

The team's electrical subject matter expert concluded that without a comprehensive inventory of all U.S. controlled facilities, it is difficult to establish an inspection plan and assess resource needs. Currently, the local commander has responsibility for repairs in his/her area. A comprehensive inventory of facilities in Afghanistan could be used by electrical inspectors and repair personnel to ensure that all U.S. controlled facilities are safe. The contractor listing can be used as a starting point for developing a complete accounting of U.S. controlled facilities in Afghanistan. However, using LOGCAP listings alone would leave at least 163 FOBs without a facility listing.

Recommendation, Client Comments, and Our Response

2. We recommended that USFOR-A identify and record all facilities controlled by U.S. forces in Afghanistan.

Client Comments to Observation and Recommendation

CENTCOM concurred with exceptions. CENTCOM stated that they tasked USFOR-A and Army Central Command to use their combined efforts to create a comprehensive database, identifying all U.S. controlled facilities in Afghanistan. CENTCOM further stated that USFOR-A Engineering would develop a single composite database compiled from all available databases in Afghanistan to meet the needs of all USFOR-A staff departments. On June 29, 2009, CENTCOM stated that all existing facility data was being consolidated into an accessible database located on the USFOR-A website.

Our Response

CENTCOM comments were responsive. However, we urge CENTCOM and USFOR-A to ensure that the databases being used to compile the comprehensive list are complete and accurate. Verification and existence/completeness testing of database records should be completed. The database needs to be kept current as new facilities are built and old facilities destroyed or removed.

Observation 3. Comprehensive Electrical Inspection Plan for U.S. Controlled Facilities in Afghanistan

We observed the lack of a comprehensive electrical inspection plan for U.S. controlled facilities in Afghanistan. The Commanders in Afghanistan were aware of the potential electrical issues and worked to address life, health, and safety issues. During our visit, there were a number of initiatives underway indicating that Commanders in Afghanistan were aware of the risks and taking steps to improve the electrical safety of facilities under their control.

For example, TF POWER was established by CJTF-101 in October of 2008, to "prevent the loss of life and government property through immediate and long-term measures that will significantly reduce the number of electrical and fire incidents throughout the combined/joint operations area." TF POWER used contractors to review and identify electrical deficiencies to include life, health, and safety issues at FOBs. According to TF POWER representatives, as of April 18, 2009, it tracked electrical inspections at 216 FOBs and completed 100% inspections of 16 bases. However, as noted earlier, without a full listing of facilities, it is difficult to determine or track whether all facilities at those 16 bases were inspected. During the time of our visit, TF POWER did not track all 257 FOBs in Afghanistan, leaving 41 FOBs at risk of not being fully reviewed for electrical safety. The number of 257 FOBs is expected to increase dramatically as the number of troops in Afghanistan increases.

In addition to the TF POWER electrical reviews, the CJTF-101 IG and Expeditionary Sustainment Command-Afghanistan IG expanded their base inspections to include electrical reviews. During one of our meetings, the CJTF-101 IG provided a presentation containing pictures of some of the electrical deficiencies that were identified and reported to the Command. Corrective actions were being tracked by the CJTF-101 IG. According to staff at FOB Sharana, there was also an annual winterization process which touches all FOBs in Afghanistan and includes some component of an electrical evaluation. However, identification and correction of electrical issues was not the primary purpose of either of these processes.

While all of the aforementioned initiatives are noteworthy initial actions; more still needs to be accomplished. A single plan for the inspection of all U.S. controlled facilities is needed. The comprehensive inventory discussed in observation 2 is an integral part of establishing an inspection plan and the two should be implemented in unison. Once a comprehensive inventory is completed, it follows that a detailed, organized plan can be established for all repairs based on the order of merit of the facilities and safety considerations of U.S. Forces.

Recommendation, Client Comments, and Our Response

3. We recommended USFOR-A develop a comprehensive plan to inspect, detect, and correct electrical deficiencies in facilities controlled by U.S. forces in Afghanistan.

Client Comments to Observation and Recommendation

CENTCOM concurred with exceptions. CENTCOM stated that a comprehensive plan was under development by TF POWER for a third party assessment and evaluation to inspect all electrical generation and distribution systems, then initiate proper corrective action for deficiencies at all USFOR-A occupied facilities in Afghanistan. The plan was to be completed by May 30, 2009. Additionally, USFOR-A was drafting a Standard Operating Procedure (SOP), which would address inspection, detection, and correction of electrical deficiencies in facilities controlled by U.S. forces. CENTCOM estimated completion of the SOP by June 15, 2009. On June 29, 2009, CENTCOM stated USFOR-A was finalizing their comprehensive plan. When finished, the plan will consist of three phases, (1) initial facility inspection and creation of an initial inspection report (currently underway), (2) analysis of the initial inspection report to determine the proper corrections, materials, and funding requirements, and (3) continued inspection of all maintenance work and new construction.

Our Response

CENTCOM comments were responsive. We request copies of the plan and SOP when completed.

Observation 4. Resources to Inspect, Detect, and Correct Electrical Deficiencies

We observed an apparent shortage in resources to properly inspect, detect, and correct electrical deficiencies. The shortages included qualified engineers, inspectors, electricians, and funds. The shortage also included qualified and available electricians to conduct the electrical work and qualified Contracting Officer's Representatives (CORs) to review contracted electrical work and administer the contracts.

At Bagram Air Field we were informed that the FOBs rely largely on Department of Defense personnel and local contractors for electrical support. Some were using civilian-run generators and most used spot generation as there were no grids. As RC-South expands, the Department will need engineering support and structure to be able to sustain the growth.

According to DCMA-Afghanistan staff, DCMA identified a shortage of qualified CORs to review contracted electrical work. DCMA concluded that they had 185 CORs; their review indicated a need for 512 CORs. DCMA estimated that they were understaffed by 327 CORs. DCMA is the quality assurance representative and the property administrator for LOGCAP contracts. This issue is discussed in greater detail in observation 11.

At Camp Brown at Kandahar Airfield we observed Seabees performing electrical work to correct noncompliance issues. At the time of our visit, there were five Seabees and one master electrician conducting the electrical work at Camp Brown. One believed that four or five senior electricians were needed to efficiently perform the electrical work on the camp.

The U.S. Army Corp of Engineers-Afghanistan Engineer District (USACE-AED) commander informed us during our visit that he was tasked with an increasingly significant construction workload involving Afghanistan reconstruction and support to a plus-up of U.S./Coalition forces. The USACE-AED commander indicated he did not have the personnel to assist in additional quality oversight of non-USACE projects.

The lack of proper personnel to inspect, detect, and correct electrical deficiencies increased the risk of undetected electrical defects, non-compliant construction and repairs being performed, and the acceptance of non-compliant electrical work performed by contractors. This increased the risk of injury, fires, and property damage.

Recommendation, Client Comments, and Our Response

4. We recommended that USFOR-A complete a comprehensive electrical inspection to determine requirements. Using those results, acquire the necessary financial and personnel resources to maintain continuous electrical inspections and corrective action program.

Client Comments to Observation and Recommendation

CENTCOM concurred. CENTCOM further stated that USFOR-A was developing a comprehensive inspection program to verify the condition of the existing electrical generation and distribution systems at all U.S. occupied facilities in Afghanistan. USFOR-A plans to use the results of that inspection program as the basis for determining the adequacy of LOGCAP and other contractor staffing plans. On June 29, 2009, CENTCOM stated that Task Force POWER continues their comprehensive inspection of existing facilities. The initial inspection team includes an organization of 116 personnel (including eight engineers, three master electricians, 72 subject matter experts/electrical inspectors, and 33 support personnel). After the initial inspection is complete, these personnel will transform into the continued inspection staff of 42 personnel. Once initial inspections are completed and results analyzed, requirements for funding, materials, and manpower will be requested.

Our Response

CENTCOM comments were responsive.

Observation 5. Level of Recordable Electrical Accidents in Afghanistan

DoD accidents are classified based on the severity of the injury, occupational illness, or property damage. A Class C accident is the lowest classification reported, and does not cover electrical shocks that do not result in a non-fatal injury or non-fatal occupational illness. Currently, a Class C accident is one that causes a non-fatal injury and causes any loss of time from work beyond the day or shift it occurred; a non-fatal occupational illness that causes loss of time from work; disability at any time; and/or property damage totaling at least $20K but less than $200K.

As such, many incidents of electrical shock were not required to be reported. Based on fieldwork conducted, we determined that personnel have experienced electrical shocks in facilities, such as the showers, which under the current policy, were not required to be reported.

According to the assessment team's electrical subject matter expert, electrical shocks traumatize the body and should be treated as near electrocutions. By maintaining a system for reporting for all electrical shocks, the Command would be better equipped to respond to electrical hazards.

Recommendation, Client Comments, and Our Response

5. We recommended that USFOR-A establish a mechanism for all electrical shocks to be reported through appropriate channels.

Client Comments to Observation and Recommendation

CENTCOM concurred. Specifically, CENTCOM reported that USFOR-A was working with all organizations, civilian and military, to collect all reports of electrical shock. A 24x7x365 telephone hotline is under development, as well as a webpage for identification of electrical problems/concerns to TF POWER. Prior to establishment of the hotline and completion of the webpage, all units have been instructed to report all electrical incidents through the chain of command for tracking and confirmation of proper resolution. On June 29, 2009, CENTCOM stated that electrical shocks were being reported through the chain of command. USFOR-A provided direction to immediately de-energize any facility where an electrical shock occurred.

Our Response

CENTCOM comments were responsive.

Observation 6. Full-Time Cadre Dedicated to Electrical Safety in U.S. Controlled Facilities in Afghanistan

During our visit, there was no full-time cadre dedicated to electrical safety of U.S. controlled facilities in Afghanistan. As discussed in observation 3, the formation of TF POWER is a great step towards ensuring electrical safety in Afghanistan. In response to electrocutions in Iraq, Task Force Safe Actions for Fire and Electricity (TF SAFE) was established by Multi-National Force-Iraq.

In its current form under CJTF 101, TF POWER does not have authority to oversee all facilities in Afghanistan. According to Command staff, TF POWER does not have a dedicated cadre of personnel to oversee the work performed by contractors. Specifically, during our visit, TF POWER was staffed by CJ-7 engineers (part time), electricians (part-time), and Prime Power (part-time). The TF POWER concept needs to be elevated to the USFOR-A level to ensure appropriate authority to oversee all facilities in Afghanistan.

According to the assessment team's electrical subject matter expert, a systematic approach, similar to that provided by TF SAFE in Iraq, is necessary to ensure that all U.S. controlled facilities in Afghanistan are electrically sound. An overarching authority is needed to direct the inspection and detection process and begin comprehensive repairs country-wide.

Recommendation, Client Comments, and Our Response

6. We recommended that USFOR-A establish, through the Joint Manning Document (JMD) process, a full-time cadre that is dedicated to electrical safety in Afghanistan.

Client Comments to Observation and Recommendation

CENTCOM concurred. CENTCOM stated that a JMD would be submitted by May 15, 2009. The JMD would include Management Oversight and appropriate technical/electrical/power experts, and would be modeled after successful programs in Iraq and Kuwait. On June 29, 2009, in response to the draft report, CENTCOM stated that USFOR-A Task Force Power had defined a JMD of 10 personnel to be added to the USFOR-A staff for electrical code enforcement and management. The submission of the JMD was delayed until completion of the 60-day assessment of the Afghanistan area of responsibility by the newly appointed USFOR-A Commander.

Our Response

CENTCOM comments were responsive. We request that CENTCOM provide us with a copy of the JMD once submitted and approved.

Observation 7. Cap on Use of Operations and Maintenance Funds for Minor Construction in Afghanistan

Multiple parties expressed that the $750,000 cap on funds for minor construction resulted in inefficient use of funds resulting in piece-mealing (splitting a construction project for power grids[2] into multiple phases) or down-scoping projects (by removing items from the statement of work).

During a meeting with key leaders in CJTF-101, they indicated that the $750,000 cap on operations and maintenance construction was not adequate for building in the high operations tempo in Afghanistan. For example, the cap dollar amount restricted efforts to create an entire power grid, which led to the construction of a series of mini-grids. The use of mini-grids can lead to higher fuel usage, additional maintenance and grounding issues resulting from "spot" generation using tactical generators.

USACE-AED personnel also stated that they support increasing the military construction cap. As a result of cap restrictions, USACE-AED indicated that they are often forced to down-scope construction activities. For example, they noted a re-locatable building project that was down-scoped by removing the fire suppression systems. CENTCOM pointed out that in FY04, the cost to construct a dining facility was $500K. Today, they indicated that it costs $2.2 million to construct a similar facility and requires congressional notification.

CENTCOM initiated a legislative proposal to have the $750,000 cap increased to $3 million for "Unspecified Minor Construction Projects." In a letter to the Office of Legislative Counsel, dated May 7, 2009, the CENTCOM commander noted that without authorization to increase the cap, "our ability to perform mission-essential military construction to support the increase of forces is at risk." He further stated in the letter that increasing the cap "will provide the resources needed to support the President's troop increase and will enhance our ability to rapidly provide essential military construction projects that will help protect, feed, care, maintain, and sustain our forces." Subsequent to us providing the Command with our preliminary observations and recommendations, the legislative proposal was submitted to Congress on May 13, 2009, as Section 1301 of the National Defense Authorization Bill for FY 2010. (See Appendix C for a copy of proposed section 1301, "Temporary Increase in Cost Threshold for Use of Operation and Maintenance Funds for Unspecified Minor Military Construction Projects in Afghanistan.")

CENTCOM provided the DoD IG a list of 21 projects that could be quickly supported if the military construction cap was raised to $3M. This list of projects included medical facilities, contingency housing, dining facilities, and electrical grids. Given the nature of

---

[2] An electrical grid is an interconnected network for delivering electricity from suppliers to consumers.

such facilities and the operational impact on U.S. forces in Afghanistan, the current legislative proposal submitted by CENTCOM should merit serious consideration.

Recommendation, Client Comments, and Our Response

7. We recommended that CENTCOM, in coordination with USFOR-A, submit a legislative proposal to increase $750,000 cap on minor construction for contingency operations in Southwest Asia.

Client Comments to Observation and Recommendation

On June 29, 2009, CENTCOM stated that Congress did not support the proposal to increase the threshold. Instead, the congressional staff recommended that DoD streamline the Contingency Construction Authorities and existing Unspecified Minor Military Construction request process to expedite construction execution.

Our Response

Although the legislative proposal was submitted on May 13, 2009, the proposed legislation was not addressed by congress as of the date of the draft report. Therefore, our recommendation remained in the draft report and CENTCOM responded to the recommendation. CENTCOM comments and actions were responsive.

Observation 8. Authority Having Jurisdiction to Grant Waivers to the National Electrical Code in Afghanistan

During meetings with staff from USFOR-A and CJTF-101, we determined that there was no arbitration cell (Authority Having Jurisdiction [AHJ]) within Afghanistan with the ability to grant waivers to the NEC. Additionally, work within Afghanistan was performed to multiple standards, to include NEC, British, and local.

The NEC was established as the electrical code for the CENTCOM area of operations by Fragmentary Order in October of 2008. According to the assessment team's subject matter expert, applying the NEC 2008 to all of Afghanistan creates difficulties. The NEC works well only if interpreted by qualified personnel. CORs did not always have the necessary qualifications to interpret the electrical codes for the work they were overseeing. The NEC, and any other electrical code, is written for technical experts in the electrical field.

According to the NEC, the AHJ is responsible for approving equipment, materials, installation, or procedures. Specifically, the AHJ is responsible for electrical inspections; investigation of electrical fires; and the design, alteration, modification, construction, maintenance, and testing of electrical systems and equipment. Additionally, the AHJ is also permitted to:

- order the immediate evacuation of any occupied building deemed unsafe when such building has hazardous conditions that present imminent danger to building occupants, and

- waive specific requirements of the NEC or permit alternative methods where it is assured that equivalent objectives can be achieved by establishing and maintaining effective safety. Technical documentation shall be submitted to the AHJ to demonstrate equivalency and that the system, method, or device is approved for the intended purpose.[3]

Although not required by the NEC 2008, the Department of Defense has a history of utilizing an AHJ in forward operating environments, which is evidenced by inclusion of an AHJ in the base LOGCAP contract and the use of an AHJ in direct support of TF SAFE in Iraq. By not having an established AHJ in Afghanistan, the Department of Defense does not have assurance that decisions being made regarding the electrical safety of facilities and installation procedures are being made in a consistent, professional manner by Department of Defense officials.

---

[3] National Fire Protection Association (2007). National Electrical Code -2008 Edition (NFPA 70).

Recommendation, Client Comments, and Our Response

8. We recommended USFOR-A establish an AHJ with the authority to review requests and grant code deviations, as necessary. The AHJ should be established at USFOR-A and in accordance with guidance in NEC 2008, Annex H.

Client Comments to Observation and Recommendation

CENTCOM concurred. The TF POWER officer in charge will be designated by USFOR-A as the AHJ in Afghanistan (Local AHJ). Issues will be submitted to the officer in charge for adjudication and policy enforcement. Additionally, requested waivers or deviations from the NEC will be submitted to the Tri Services Electrical Working Group for review and consideration. On June 29, 2009, CENTCOM stated that an AHJ had been established and is responsible for all electrical issue adjudication, policy development, and code enforcement.

Our Response

CENTCOM comments were responsive.

Observation 9. Training Soldiers on Electrical Hazards and the Reporting Process

A comprehensive program did not exist to educate and increase awareness of electrical safety for soldiers prior to arriving in theater. Further, there was a lack of in-theater training for individual soldiers on awareness of electrical hazards, the reporting process, and consequences for tampering with electrical circuits. According to the safety officers, many soldiers made electrical repairs on their own, and they were not trained electricians qualified to work on electrical systems.

Training is essential to ensure that soldiers know how to avoid electrical hazards, such as shocks, are informed on reporting procedures, and aware of the consequences of tampering with electrical systems or components.

Without such training, soldiers may be at risk of electrical shocks or injury to themselves or others and may also risk potential property damage. Further, incidents of electrical shock that are not reported may not be properly addressed and may result in injury to additional soldiers.

Recommendation, Client Comments, and Our Response

9. We recommended that USFOR-A include training on electrical safety and incident reporting as part of pre-deployment and in-theater training. Training should also emphasize consequences for tampering with electrical circuits.

Client Comments to Observation and Recommendation

CENTCOM concurred. USFOR-A is developing an electrical safety training program to include PowerPoint slides that will be briefed to personnel during Reception, Staging, Onward-movement, and Integration. A standard operating procedure is under development, which will emphasize the importance of electrical safety within the workplace and living spaces, reinforcing the role of the senior non-commissioned officers in troop leadership. Units will have a safety organization that includes fire and electrical inspections to reinforce awareness of electrical safety and hazards. On June 29, 2009, CENTCOM stated that TF POWER has developed training slides to be added to the Theater RSO&I briefing program. Further, training of individual unit Training Officers will be implemented upon arrival of the USFOR-A Safety Officer and Staff after submission to fill the JMD.

Our Response

CENTCOM comments were responsive. We request that USFOR-A provide us with the training program and standard operating procedures upon completion.

Observation 10. Use of Unlisted Electrical Components

CJTF-101 identified the use of unlisted power strips and transformers[4] by soldiers. They stated that the unlisted power strips and transformers were purchased in the post exchanges and from other vendors. CJTF-101 has since issued standards on electrical components sold on CJTF-101 installations. However, although many of them had been confiscated during inspections, not all had been collected and may have still been in use by soldiers and civilians. Further, some of the power strips and transformers that were confiscated were not replaced with those that were certified and authorized for use because of the unavailability of listed replacements.

According to the assessment team's subject matter expert, unlisted power strips did not have circuit breakers built in, which allows for the free flow of current beyond the nominal rating of the power strip. This may cause electrical shock and fires.

Criteria

Memorandum, "Standards for Electrical Components Sold on CJTF-101 Installations," April 6, 2009, establishes standards for electrical power strips sold by vendors authorized to conduct business on military installations within CJTF-101. It prohibits the sale of power strips by vendors other than the Army Air Force Exchange System Post Exchange. This policy ensures that soldiers and civilians supporting Operation Enduring Freedom are protected from electrical hazards in their living areas and workplaces.

Recommendation, Client Comments, and Our Response

10. We recommended that USFOR-A establish a program to provide listed replacement power strips and transformers.

Client Comments to Observation and Recommendation

CENTCOM concurred. CENTCOM further stated that TF POWER will coordinate the ordering of approved power strips and transformers for the unit and command safety officers to exchange and replace the unlisted units.

Our Response

CENTCOM comments were responsive.

---

[4] Listed power strips and transformers are those that have been certified by the Underwriters Laboratory (UL), Canadian Standards Association (CSA), or Conformance European (CE).

Observation 11. Qualified Contracting Officer's Representatives for Review of Electrical Work

A lack of theater personnel qualified to perform as CORs was identified by the DCMA as a challenge throughout Afghanistan. Additionally, good business practices require the separation of duties in contract execution; the lack of qualified CORs places the separation of duties principle at risk. Because of the technical requirements needed to oversee electrical contracts, the problem was particularly acute.

According to DCMA, they had recently performed a requirements review for CORs in Afghanistan. At the time of their review, they had 185 CORs and the review indicated a need for a total of 512 CORs. Therefore, they indicated they were understaffed by 327 CORs total, 37 short in the fields of power generation and electrical distribution. They also stated that the number of COR audits received was less than one per COR per month; the requirement for most CORs was four reports per month. Electrical issues were considered high risk for DCMA quality assurance reviews.

DCMA indicated that they were attempting to garner additional resources to solve the COR shortage. They were working with USFOR-A CJ7 for theater engineer resources to serve as CORs and were coordinating with the 249th Engineer Battalion to leverage theater electrical subject matter experts to serve as CORs. Additionally, they indicated that they were working with CJTF-101 to complete an official "gap analysis" to best determine the exact number of necessary personnel.

DCMA also stated that they had limited technical subject matter experts to provide Government oversight for LOGCAP high risk services, primarily in facilities management, construction, engineering services, power distribution, and power generation.

During a meeting with the 249th Engineer Battalion, personnel noted that TF SAFE had identified 385 soldiers in Iraq with electrical military occupational specialties. They explained that these individuals were not typically assigned jobs related to monitoring or maintaining low voltage systems. The same may hold true for Afghanistan; if so, these types of soldiers could be used as CORs.

A strong cadre of CORs would mitigate risk of unchecked contractor performed services throughout Afghanistan. A plus-up of CORs would allow senior leadership at TF POWER to replace existing contractor personnel performing inspections on contractor work and alleviate a concern regarding separation of duties.

Recommendation, Client Comments, and Our Response

11. We recommended that USFOR-A, in coordination with DCMA, identify and train the individuals needed to satisfy COR requirements.

Client Comments to Observation and Recommendation

U.S. Central Command Comments

CENTCOM concurred with exceptions. CENTCM stated that DCMA had an aggressive COR expansion plan to increase in theater electrical system CORs from 16 to 53. Additionally, CENTCOM stated that USFOR-A TF POWER, in coordination with DCMA will create an organizational chart which clearly explains the relationships between the COR and the worker installing the work. On June 29, 2009, CENTCOM stated that the anticipated arrival date in theater of USACE architecture and engineering subject matter experts was September 16, 2009. Further, the A&E subject matter experts were to be trained as Contracting Officer's Technical Representatives to collect inspection reports and develop the proper documentation for the responsible COR.

Defense Contract Management Agency Comments

On July 9, 2009, DCMA concurred with our finding and recommendation. DCMA also provided additional information regarding the COR training process which included both formal and on-the-job training.

Our Response

CENTCOM and DCMA comments were responsive. We request CENTCOM provide us with a copy of the organizational chart once completed.

Observation 12. Re-wiring of New Ablution Units at Kandahar Air Field

We observed that new ablution units at Kandahar Airfield were being re-wired to meet NEC 2008. The assessment team's subject matter expert concluded that there may be a timelier and more cost effective alternative to meet NEC other than a complete re-wiring. In the opinion of the subject matter expert, replacing the electrical panel in lieu of complete rewiring may be a solution that could bring the Ablution units into compliance with NEC 2008.

Ablution units are re-locatable buildings which were being used by U.S. and North Atlantic Treaty Organization (NATO) forces. A typical ablution unit contained five sinks and five showers. These units were purchased prefabricated (wiring and plumbing are complete upon delivery). Because these were new buildings intended for occupancy by U.S. forces, they should be NEC 2008 compliant.

During our walk through of Kandahar Airfield, we met with a representative from the NATO Maintenance and Supply Agency who informed us that the units pictured below were being rewired because they were not NEC 2008 compliant. He also stated that these units were part of an order of 150 brand new buildings that would cost the Department of Defense over $600,000 to rewire.

Figure 14. Outside view of a typical ablution unit

Figure 15. Inside view of a typical ablution unit

According to the assessment team's electrical subject matter expert, these units were wired using Residual Current Device, the European equivalent of ground fault current interrupters (GFCI). Due to a lack of understanding of NEC requirements and how codes work together, the COR instructed the contractor to rewire all ablution units "to meet the code." A qualified electrician should be made available to inspect the units and determine if a less costly modification can be made to meet the specifications of the NEC 2008.

There was no AHJ to make code determinations, as such, the COR was left to make a technical decision regarding electrical safety and sufficiency. This could result in a more expensive fix than is necessary to meet the intent of the NEC 2008.

Recommendation, Client Comments, and Our Response

12. We recommended that USFOR-A determine if replacing the electrical panel, in lieu of complete re-wiring, will bring the ablution units in compliance with NEC 2008.

Client Comments to Observation and Recommendation

CENTCOM concurred. Specifically, CENTCOM stated that there were two issues being corrected at Kandahar Air Field. First, the initial deliveries of ablution units from Corimec were wired using plastic electrical panels and non-GFCI receptacles over the sinks. This required the upgrade of the electrical panel and receptacles in 36 Corimec units. Corimec was made aware of the NEC requirement and all remaining units were to be corrected at the factory. Second, 32 units were received from a local supplier. The electrical wiring in these locally supplied ablution units were below acceptable standards, and all require a complete electrical wiring replacement. The local supplier was no longer being used. On June 29, 2009, CENTCOM stated that the changes to existing containers were necessary, the contractor was being responsive in correcting this issue during manufacturing, and that correction of the discrepancies continued.

Our Response

CENTCOM comments were responsive. We request CENTCOM provide us with an update on status of the corrective actions underway.

Observation 13. Kandahar Air Field Power Plant

According to the site manager, a contractor with IAP Worldwide Services, the power plant at Kandahar Air Field was at maximum capacity (22 MW) and usage was expected to expand by 75 percent by next year. Problems included:

- Last year, peak demand was 20 MW and was expected to increase by 15 MW by next year.
- At the time of our visit, the power plant location would allow for only an additional generation of 12 MW.
- Power shortages may require spot generation.

According to the contractor, a plan was in place to open an additional power plant at an additional location. Further, the site manager informed the team that this was a NATO-managed power plant. The USFOR-A personnel had no input or control over this plant, the operation, maintenance or required upgrades to the plant.

We are highlighting this issue for context issues regarding electrical capacity in Afghanistan.

Figure 16. Kandahar Power Plant

Client Comments, and Our Response

Client Comments to Observation
CENTCOM concurred with the observation.

Our Response
CENTCOM comments were responsive.

Appendix A. Scope and Methodology

We conducted this assessment from March 31, 2009, through May 29, 2009, in accordance with Quality Standards for Federal Offices of Inspector General, and visited sites in Afghanistan from April 19, 2009, to April 25, 2009. We planned and performed the assessment to obtain sufficient and appropriate evidence to provide a reasonable basis for our observations, conclusions, and recommendations, based on our assessment objective.

The scope of our assessment encompasses the electrical safety of Department of Defense-occupied and -constructed facilities in Afghanistan during the period of January 2008, to present. Specifically, the assessment team was to determine the effectiveness of command efforts to ensure the electrical safety of U.S. Military, civilians, and contractor personnel in Defense-occupied and -constructed facilities in Afghanistan. We conducted interviews with Commands, agencies, and contractors at the following site visits in Afghanistan:

Bagram Air Field:
- Task Force Protecting Our Warfighters and Energy Resources (TF POWER);
- Defense Contract Management Agency (DCMA) – Afghanistan;
- Joint Contracting Command-Afghanistan – Principal Assistant Responsible for Contracting;
- Logistics Civil Augmentation Program (LOGCAP) – Afghanistan;
- Defense Logistics Agency;
- Combined Joint Task Force (CJTF)-101;
- Inglett & Stubbs International;
- Fluor Corporation; and
- KBR.

Kandahar Air Field:
- U.S. Forces-Afghanistan (USFOR-A);
- TF POWER;
- TF Anzio;
- 143rd Expeditionary Sustainment Command IG;
- DCMA-Afghanistan;
- North Atlantic Treaty Organization Maintenance and Supply Agency;
- Fluor Corporation; and
- KBR.

Kabul, Afghanistan:
- USFOR-A;
- USFOR-A, Engineers
- Combined Security Transition Command-Afghanistan;
- U.S. Army Corps of Engineers-Afghanistan Engineering District;
- DCMA-Afghanistan;
- LOGCAP; and
- KBR.

We also contacted CENTCOM to discuss the observations and recommendations, as well as visited the 249th Engineer Battalion at Ft. Belvoir, VA and Forward Operating Bases Altimur and Sharana (Bagram) and Spin Boldak and Tarin Kowt (Kandahar).

We reviewed documents such as Army regulations and pamphlets on facilities management and engineering, DoD instructions on a safety and occupational health program and accident investigation, reporting, and recordkeeping. We also reviewed policies, task orders, fragmentary orders, and inspection reports on the electrical safety of DoD Military and civilians, as well as electrical construction and repairs of DoD-controlled facilities.

Limitations

We limited our review to the current status of efforts made by the Department of Defense personnel, contractors, and Afghan local nationals to assess and repair electrical deficiencies in U.S.-controlled facilities. The team's subject matter expert (senior electrician from A Co., 249th Engineer Battalion (Prime Power)) performed limited assessments of electrical components at select facilities in Afghanistan. The subject matter expert did not conduct full electrical inspections of each facility due to time constraints in country. Based on these assessments observations were made on the conditions of electrical components, wiring, panels, bonding, and grounding, in order to form conclusions on electrical safety.

Methodology

On April 19, 2009, we began a one week assessment of the electrical safety of DoD facilities occupied or constructed by U.S. personnel and contractors in Afghanistan. The assessment team performed work at Bagram Airfield (BAF), Forward Operating Base (FOB) Altimur, FOB Sharana, Kandahar Airfield (KAF), Camp Brown, FOB Tarin Kowt, FOB Spin Boldak, Camp Phoenix, and Camp Eggers.

We used a subject matter expert to assess the electrical safety and code compliance of facilities visited at various locations. The conclusions in the report are based on limited reviews, first-hand observations, and interviews.

We interviewed senior military leaders at U.S. Forces – Afghanistan; Combined Security Transition Command – Afghanistan; and Combined Joint Task Force-101.

Use of Computer-Processed Data

We did not use any computer-processed data in this assessment.

Use of Technical Assistance

The assessment team was augmented by a subject matter expert, a senior electrician from A Co., 249th Engineer Battalion (Prime Power), to conduct limited reviews of the electrical safety and code compliance of facilities visited at various locations in Afghanistan.

Acronyms Used in this Report

The following is a list of the acronyms used in this report.

AHJ	Authority Having Jurisdiction
ARCENT	U.S. Army Forces, U.S. Central Command
BAF	Bagram Air Field
CENTCOM	U.S. Central Command
CJTF	Combined Joint Task Force
COR	Contracting Officer's Representative
DCMA	Defense Contract Management Agency
FOB	Forward Operating Base
GFCI	Ground Fault Current Interrupter
JMD	Joint Manning Document
KAF	Kandahar Air Field
LOGCAP	Logistics Civil Augmentation Program
NATO	North Atlantic Treaty Organization
NEC	National Electrical Code
O&M	Operations & Maintenance
OIG	Office of the Inspector General
RC	Regional Command
RSOI	Reception, Staging, Onward movement, and Integration
SOP	Standard Operating Procedure
TF POWER	Task Force Protecting Our Warfighters and Energy Resources
TF SAFE	Task Force Safe Actions for Fire and Electricity
USACE-AED	U.S. Army Corp of Engineers-Afghanistan Engineer District
USFOR-A	U.S. Forces-Afghanistan

Appendix B. Additional Photographs

This appendix contains additional photos taken by the OIG assessment team during the various Afghanistan FOB and camp tours. The DoDIG assessment team was augmented by an electrical subject matter expert. The captions represent opinions of the subject matter expert.

Figure B-1

Figure B-2

Figures B-1 and B-2 were taken at FOB Altimur and show improperly grounded generators. Figure B-2 (which is a close-up photo of figure B-1) clearly shows multiple ground rods tied together and ground rods that are not at least 6 feet apart (as required by the National Electrical Code).

Appendix B. Additional Photographs (continued)

Figure B-3

Figure B-4

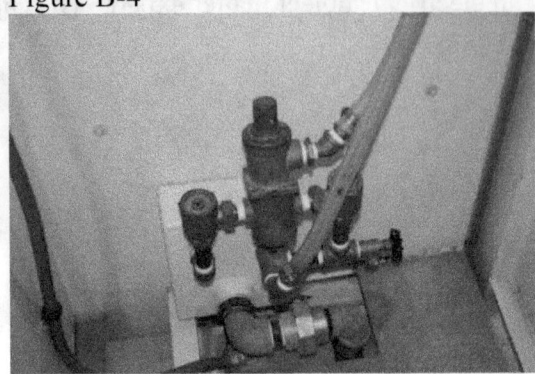

Figure B-5

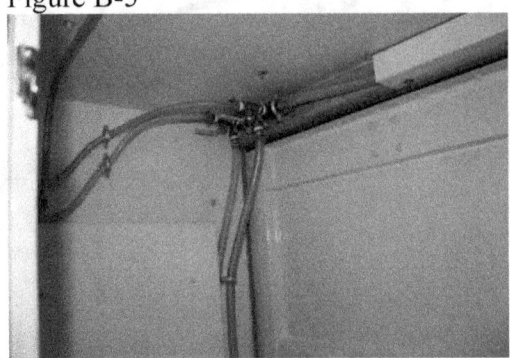

Figures B-3, B-4, and B-5, were taken at FOB Altimur. The risk of shock is reduced by using rubber or PVC pipes to interrupt or in place of the usual copper tubing in shower units. The rubber (or PVC) piping does not conduct electricity. As such, it interrupts the electrical circuit. This means if a short were to form in the system, the risk of the water becoming electrified is greatly reduced.

Appendix B. Additional Photographs (continued)

Figure B-6

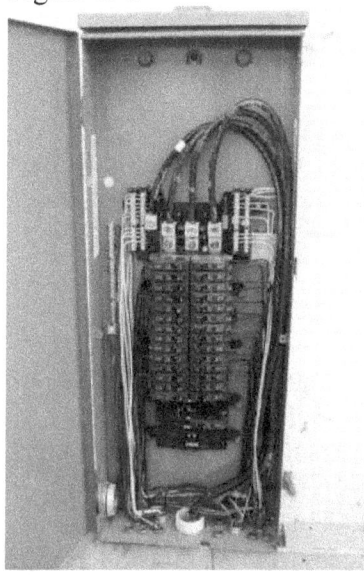

Figure B-7

Figures B-6, and B-7 are pictures of electrical panels at FOB Altimur. Figure B-6 shows a clean, grounded, and bonded box. Figure B-6 also shows a cost effective alternative to rewiring an entire facility when the wires are not color coded to NEC standard. Specifically, the use of colored tape on the ends of corresponding wires or a permanent marker such as paint can be used to bring the box in line with the NEC. Additionally, Figure B-6 shows a wood panel underneath the box that is being used as a temporary fix to protect exposed wires leading into the box. While still a fire hazard and not code compliant, the wooden box temporarily keeps personnel from contacting live wires. Although not pictured, the panel in figure B-6 also had a dead front. Figure B-7 shows a box that did not have the properly colored coded wires; however this box did have a listing of the circuits (panel schedule).

Appendix B. Additional Photographs (continued)

Figure B-8

Figure B-9

Figures B-8 and B-9 were taken of tents being used to house personnel at Kandahar Air Field. The Harvest Falcon Package is a prefab tent that holds 36 troops and comes with power strips and lighting as part of the prefabricated package. The power strip shown in figure B-9, should have been run out of the way of high traffic areas to avoid trip hazards. Additionally, the limited number of outlets for 36 personnel might encourage daisy chaining; adding additional outlets could decrease safety and electrical hazards.

Figure B-10

Figure B-10 was taken at Camp Eggers. Specifically, this was taken at a Morale, Welfare, and Recreation center on Camp Eggers. The photo is of the electrical panel which had a dead front and a listing of the circuits (panel schedule).

Appendix B. Additional Photographs (continued)

Figure B-11.

Figure B-11 was taken at Camp Eggers. This is a photo of an overhead raceway for cables to protect them from getting stepped on or run over.

Figure B-12

Figure B-12 was taken outside of a barracks at Camp Eggers. The photo shows power strips rated for indoor use being used outside.

Appendix B. Additional Photographs (continued)

Figure B-13   Figure B-14

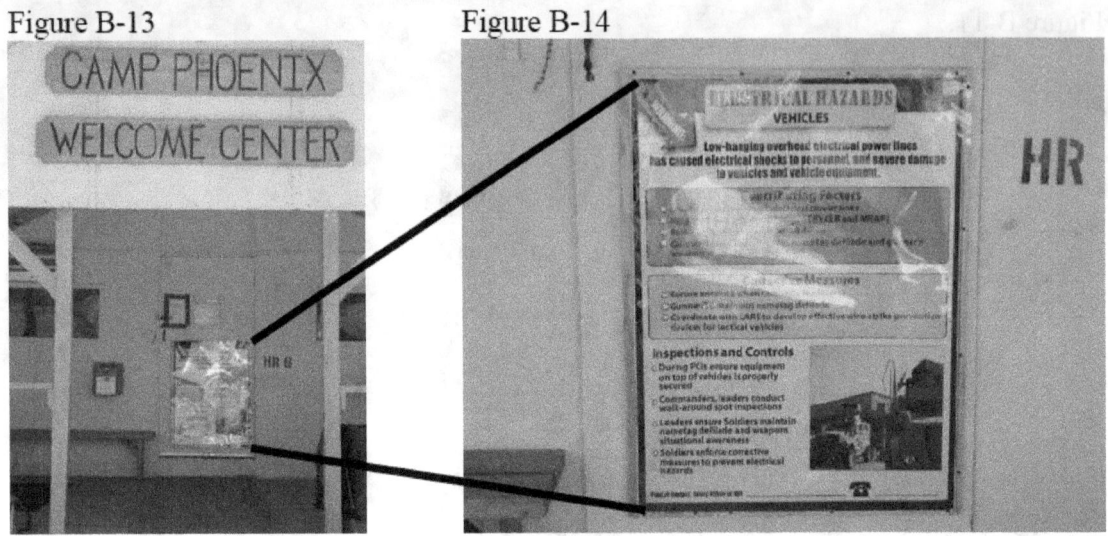

Figures B-13 and B-14 were taken of the Welcome Center at Camp Phoenix. Figure B-14 (which is a close-up photo of figure B-13) depicts a poster containing information on electrical safety and hazards. Specifically, the poster discuses electrical hazards related to low-hanging overhead electrical power lines.

Appendix C. Legislative Proposal Concerning Minor Military Construction Projects

TITLE XIII—MILITARY CONSTRUCTION, MILITARY FAMILY HOUSING, AND REAL PROPERTY

SEC. 1301. TEMPORARY INCREASE IN COST THRESHOLD FOR USE OF OPERATION AND MAINTENANCE FUNDS FOR UNSPECIFIED MINOR MILITARY CONSTRUCTION PROJECTS IN AFGHANISTAN.

(a) INCREASED COST THRESHOLD.—

(1) PROJECTS IN AFGHANISTAN.—For purposes of any military construction project to be carried out in Afghanistan that is a contingency operational requirements project, the cost limitation applicable to the project under subsection (c) of section 2805 of title 10, United States Code, shall be treated as being $3,000,000, notwithstanding any otherwise applicable cost limitation under that section.

(2) CONTINGENCY OPERATIONAL REQUIREMENTS PROJECT.—For purposes of this subsection, the term "contingency operational requirements project" means a project that is necessary to meet military operational requirements involving the use of the Armed Forces in support of—

(A) a declaration of war;

Appendix C. Legislative Proposal Concerning Minor Military Construction Projects (continued)

55

1 (B) a declaration by the President of a national emergency under section
2 201 of the National Emergencies Act (50 U.S.C. 1621); or
3 (C) a contingency operation (as such term is defined in section 101 of title
4 10, United States Code).
5 (b) TRANSPARENCY.—
6 (1) QUARTERLY REPORTS.—Not later than 60 days after the end of each fiscal-
7 year quarter during which subsection (a) is in effect, the Secretary concerned shall submit
8 to the congressional committees specified in paragraph (2) a report on contracts entered
9 into during that quarter under section 2805(c) of title 10, United States Code, for which
10 the applicable cost limitation was the limitation specified in subsection (a). Each such
11 report shall include a statement of obligations during that quarter for each such contract.
12 (2) CONGRESSIONAL COMMITTEES.—The congressional committees referred to in
13 this paragraph are the following:
14 (A) The Committee on Armed Services and the Subcommittee on Defense
15 and the Subcommittee on Military Construction, Veterans Affairs, and Related
16 Agencies of the Committee on Appropriations of the Senate.
17 (B) The Committee on Armed Services and the Subcommittee on Defense
18 and the Subcommittee on Military Construction, Veterans Affairs, and Related
19 Agencies of the Committee on Appropriations of the House of Representatives.
20 (3) SECRETARY CONCERNED.—For purposes of this subsection, the term
21 "Secretary concerned" has the meaning applicable to such term under section 2805 of
22 title 10, United States Code.

Appendix C. Legislative Proposal Concerning Minor Military Construction Projects (continued)

1 (c) EXPIRATION.—No funds may be obligated for a project by reason of the provisions of
2 subsection (a) after September 30, 2011.

Appendix C. Legislative Proposal Concerning Minor Military Construction Projects (continued)

Section by Section Analysis

TITLE XIII—MILITARY CONSTRUCTION, MILITARY FAMILY HOUSING, AND REAL PROPERTY

Section 1301 would raise the spending limit on the use of operation and maintenance (O&M) funds for unspecified minor military construction (MILCON) from $750,000 to $3,000,000 when applied to projects in Afghanistan that are necessary to meet military operational requirements involving the use of the Armed Forces in support of contingency operations. Increasing the authority in Afghanistan facilitates greater flexibility in military construction by allowing O&M funds to be used for a wider variety of minor projects, such as temporary dining facilities, berthing facilities, ammunition storage, utilities, and medical facilities that are no longer affordable under the limit originally set in 2004. Given the nature of the ongoing overseas contingency operations, waiting for authorization and appropriation of a military construction project does not effectively meet operational requirements. For smaller scale projects that exceed the existing $750,000 limit, the commander often must either cancel or build to a lesser requirement in order to appropriately fund the project with O&M funds.

The proposal is temporary, expiring at the end of fiscal year (FY) 2011. The proposal supports transparency by requiring the military departments to report quarterly on their use of the higher threshold.

Rising construction costs:
Rising construction costs affect every project considered for Afghanistan. The Army component of Central Command (ARCENT) experienced annual OEF/OIF construction cost growth ranging between 20-40% per year, pushing many minor yet essential projects beyond either O&M construction thresholds or unspecified minor military construction thresholds. The following examples demonstrate this cost growth (all amounts in thousands of dollars):

Project	Location	FY04	FY08-10
Landfill	Iraq	$ 420	$ 880
DFAC	Dwyer	$ 500	$2,000
Medical Facility	Dwyer	$ 800	$2,000
Dining Facility	Wolverine	$ 880	$2,200
MedLog Warehouse	Bagram	$ 700	$3,350
Brigade HQ Building	Kandahar	$ 750	$3,500

The FY 2004 costs for similar facilities were derived from previously approved Operation and Maintenance, Army (OMA) projects built in the USCENTCOM Area of Responsibility (AOR). Although the last two example projects exceed the $3.0 million authority requested, they serve to illustrate the overall cost growth trends.

Appendix C. Legislative Proposal Concerning Minor Military Construction Projects (continued)

Section by Section Analysis

Rising cost factors include:

1) Increases in the cost of construction materials (concrete costs $500 per cubic yard (CY) in Afghanistan vs $100 per CY stateside),

2) Lack of a skilled local labor force requires outsourcing more technicians,

3) Contractors in country who cannot support the scale of construction requires going outside of the country or using larger contractors which can drive costs up due to minimal competition, and

4) Cost to ship and transport materials due to limitations in air, sea, and land access.

Impact: A $3,000,000 O&M threshold for unspecified minor military construction provides the maneuver commander in the counterinsurgency fight the flexibility to reposition forces in response to the enemy situation without delay. The threshold would permit rapid construction of enabling facilities to include force protection, medical, and temporary logistic support (e.g., fuel and ammunition storage).

Changes to Existing Law: This section would make no changes to the text of existing law.

Appendix D. Management Comments to Preliminary Observations

11 MAY 2009

SUBJECT: CENTCOM PRELIMINARY RESPONSE TO DoD IG ASSESSMENT OF ELECTRICAL SAFETY IN AFGHANISTAN.

Ref: USFOR-A'S WAY AHEAD TO IG ASSESSMENT OF ELECTRICAL SAFETY IN AFGHANISTAN dtd 6 May 2009

1. **Observation:** We observed electrical issues at Camp Brown and FOB Spin Boldak involving grounding, bonding, circuit protection, and personnel protection. Those conditions need immediate correction or will likely result in significant safety issues.

   Recommendation: Take immediate action to correct the electrical deficiencies at both locations.

   USFOR-A Way Ahead:
   Camp Brown (on Kandahar): 3 Seabees from NMCB 11 have been working to correct deficiencies at Camp Brown. ECD for work completion of major discrepancies is 30 MAY 09. Transition to Inglett and Stubbs International (ISI) or LOGCAP is under review for 1 JUN 09 to continue the remediation of electrical deficiencies.

   Spin Boldak: Spin Boldak is an ISI location. ISI will be making corrections to deficiencies at this location. ECD for all electrical work completion is 30 MAY 09, with the critical Health and Life Safety completed immediately upon arrival at the site.

   CENTCOM Response: CENTCOM concurs. Per USFOR-A SITREP received 11 May 2009, NMCB 11 has corrected the Life Safety deficiencies.

2. **Observation:** There is a lack of a comprehensive inventory of U.S. controlled facilities in Afghanistan.

   Recommendation: Identify and record all facilities controlled by U.S. forces in Afghanistan.

   USFOR-A Way Ahead: USFOR-A ENG will develop a single composite database compiled from all available databases in Afghanistan. This is a collaborative effort to meet the needs of all USFOR-A staff departments.

   CENTCOM Response: CENTCOM concurs with exceptions. CENTCOM has tasked USFOR-A and ARCENT, via Tasker RFI 20090511-015 to use their combined efforts to create a comprehensive data base, identifying all U.S. controlled facilities in Afghanistan.

Appendix D. Management Comments to Preliminary Observations (continued)

UNCLASSIFIED

3. **Observation:** There is a lack of a comprehensive electrical inspection plan for U.S. controlled facilities in Afghanistan.

    Recommendation: Develop a comprehensive plan to inspect, detect, and correct electrical deficiencies in facilities controlled by U.S. forces in Afghanistan.

    USFOR-A Way Ahead: A comprehensive plan is under development by TF POWER for a third party A&E team to inspect all electrical generation and distribution systems, then initiate proper corrective action for all deficiencies at all USFOR-A occupied facilities in Afghanistan. The plan will be completed by 30 MAY 09 and will include the A&E inspection teams, DCMA, and LOGCAP requirements. USFOR-A is drafting a Standard Operating Procedure (SOP) which addresses inspection, detection, and correction of electrical deficiencies in facilities controlled by U.S. forces, as well as delineates the responsibilities of the Engineers and the occupants. Completion of SOP - 15 JUN 09. Inspection services will be contracted through USACE existing A&E contracts using the Task Order process. These A&E inspectors will be dispersed across Afghanistan to accomplish the required inspections. Statement Of Work (SOW) completion date is 15 JUN 09. USACE provided electrical inspectors will provide the government required QA upon arrival in theater.

    Interim action: Currently, 9 teams from ISI are conducting inspections. An additional 16 electrical repair teams will be stood up and in country between 18 MAY and 30 JUN 2009. These personnel will correct electrical deficiencies at the remote sites that have been inspected with corrective scope approved and directed by specific Task Orders (TOs) provided to ISI. An additional 30 inspection teams from ISI are under consideration. ISI has provided a cost proposal that will be processed by USFOR-A (JARB) and then ARCENT (Super CARB) to expedite the inspection schedule of the outlying FOBs. Additionally, KBR is providing QC by inspecting and correcting all electrical systems and components at KBR supported locations, verifying compliance with the 2005 NEC.

    CENTCOM Response: CENTCOM concurs with exceptions. USFOR-A is correct, with one exception. All electrical work completed prior to the release of the 2008 NEC should be inspected under the 2005 NEC. All future new construction should be conducted under the most current NEC. Contracts with ISI and KBA should be written with respect to the most current NEC.

4. **Observation:** There is an apparent shortage in resources to properly inspect, detect, and correct electrical deficiencies. The shortages include qualified engineers, inspectors, electricians, and funds.

    Recommendation: Complete a comprehensive electrical inspection to determine requirements. Using those results, acquire the necessary financial and personnel resources to accomplish the mission.

    USFOR-A Way Ahead: USFOR-A is developing a comprehensive inspection program to verify the condition of the existing electrical generation and distribution systems at all US

UNCLASSIFIED

Appendix D. Management Comments to Preliminary Observations (continued)

occupied facilities in Afghanistan. The results of that inspection program will be the basis for determining the adequacy of LOGCAP and other contractor staffing plans. Discrepancy lists created during this inspection process will be the basis for corrective projects. Established project development and funding process will be used for extensive repair projects. Maintenance efforts will be initiated by the local trouble desk system and tracked to completion for proper response and complete corrective action. Trends will be tracked and, should adequate resources be available from the contracted maintenance provider, formal requests will be made through LOGCAP and DCMA to correct the deficiency. If adequate resources are not available from the contracted maintenance provider then a SOW will be developed and separate contract will be issued for the corrective action.

CENTCOM Response: CENTCOM concurs.

5. **Observation:** Currently, the lowest level of recordable electrical accident is one that causes a non-fatal injury and causes any loss of time from work beyond the day or shift it occurred; a non-fatal occupational illness that causes loss of time from work; or disability at any time and/or property damage totaling at least $20K but less than $200K.

   Recommendation: All electrical shocks should be reported through appropriate channels.

   USFOR-A Way Ahead: USFOR-A is working with all organizations, civilian and military to collect all reports of electrical shock. A 24x7x365 telephone hot line is under development, as well as a Webpage for identification of electrical problems/concerns to TF POWER.

   Immediate response: Life and Health Safety issues, and non-Life and Health Safety issues will be corrected by the established process and by the contracted maintenance organization. Prior to establishment of the Hotline and completion of the Webpage, all units have been instructed to report ALL electrical incidents up through their chain of command for tracking and confirmation of proper resolution.

   CENTCOM Response: CENTCOM concurs.

6. **Observation:** There is no full-time cadre that is dedicated to electrical safety of U.S. controlled facilities in Afghanistan.

   Recommendation: USFOR-A needs to establish, through the Joint Manning Document process, a full-time cadre that is dedicated to electrical safety in Afghanistan.

   USFOR-A Way Ahead: A JMD will be submitted no later than 15 MAY 09 after the TF POWER program objectives, resource analysis, status and way ahead have been outlined and approved. The JMD will include Management Oversight and appropriate technical/electrical/power experts, and will be modeled after successful programs in Iraq and Kuwait.

   CENTCOM Response: CENTCOM concurs.

Appendix D. Management Comments to Preliminary Observations (continued)

UNCLASSIFIED

7. **Observation:** Multiple parties have expressed that $750,000 cap on funds for minor construction results in inefficient use of funds resulting in piece-mealing or down-scoping projects.

   Recommendation: CENTCOM, in coordination with USFOR-A, submit a legislative proposal to increase $750,000 cap on minor construction for contingency operations in Southwest Asia.

   USFOR-A Way Ahead: Working with CENTCOM to provide necessary documentation to support the legislative proposal.

   CENTCOM Response: CENTCOM is working with legal, J8, OSD and Legislative Affairs have rewritten the congressional language that will increase the $750,000 to $3,000,000. Additionally, CENTCOM has a support letter signed by GEN Petraeus supporting this increase. Since this proposal is a law, it will need to be approved by the administration and then passed by the House and Senate

8. **Observation:** There is no arbitration cell (Authority Having Jurisdiction [AHJ]) at USFOR-A with the ability to grant waivers to the NEC 2008 in Afghanistan.

   Recommendation: Establish an AHJ with the authority to review requests and grant code deviations, as necessary. The AHJ should be established at USFOR-A and in accordance with guidance in NEC 2008, Annex H.

   USFOR-A Way Ahead: CAPT Kuellmer, Task Force POWER OIC, will be designated by USFOR-A as the AHJ in Afghanistan (Local AHJ (LAHJ)). AHJ is defined as the organization, office, or individual responsible for approving equipment, materials, an installation, or a procedure. The Tri Services Electrical Working Group (TSEWG) has mandated certain limitations on the LAHJ authority that will be adhered to by the appointee. In accomplishing the responsibilities of LAHJ, USACE A&E Electrical Inspectors and support team will be on site inspectors. Issues will be submitted to CAPT Kuellmer for adjudication and policy enforcement. Any requested waivers or deviations from the NEC will be submitted to the TSWEG for review and consideration in accordance with the precepts of the TSEWG.

   CENTCOM Response: CENTCOM concurs.

9. **Observation:** There is a lack of training for individual soldiers on the awareness of electrical hazards, the reporting process, and consequences for tampering with electrical circuits.

   Recommendation: Include training on electrical safety and incident reporting as part of pre-deployment and in-theater training. Training should also emphasize consequences for tampering with electrical circuits.

UNCLASSIFIED

Appendix D. Management Comments to Preliminary Observations (continued)

UNCLASSIFIED

USFOR-A Way Ahead: USFOR-A is developing an electrical safety training program to include PowerPoint slides that will be briefed to personnel during RSOI. The SOP under development will emphasize the importance of Electrical Safety within the workplace and living spaces, reinforcing the role of the senior NCOs in troop leadership. Units will have a Safety Organization that includes Fire and Electrical inspections to reinforce awareness of electrical safety and hazards.

CENTCOM Response: CENTCOM concurs.

10. **Observation:** Command has identified the use of unlisted power strips and transformers.

    Recommendation: Establish a program to provide listed replacement power strips and transformers.

    USFOR-A Way Ahead: Unlisted power strips and transformers, have been collected at the unit level and exchanged with UL or CE rated power strips. TF POWER will coordinate the ordering of approved power strips and transformers for the unit and command Safety Officers to exchange and replace the unlisted units.

    CENTCOM Response: CENTCOM concurs.

11. **Observation:** A lack of qualified Contracting Officer Representatives (CORs) has been identified as a problem throughout Afghanistan. Additionally, good business practices require the separation of duties in contract execution; the lack of qualified CORs places the separation of duties principle at risk. Because of the technical requirements needed to oversee electrical contracts, the problem is particularly acute. We believe a potential remedy could be to identify individuals having technical skills and train them to be Contracting Officer Technical Representatives (COTRs) to support existing CORs.

    Recommendation: Identify and train the individuals needed to satisfy COTR requirements.

    USFOR-A Way Ahead: TF POWER structure is based upon a model that provides the technical skills required to inspect electrical systems and identify deficiencies. These will be reported to the DCMA CORs responsible for documentation to the responsible maintenance contractor for proper correction. Personnel will be trained as COTRs to collect the inspection reports and develop the proper documentation for the responsible COR. DCMA has an aggressive COR expansion plan to increase in theater electrical system CORs from 16 to 53.

    CENTCOM Response: CENTCOM concurs with exceptions. USFOR-A TF Power, in coordination with DCMA will create an organizational chart which clearly explains the relationships between the COR, COTR's and the worker installing the work.

12. **Observation:** New ablution units in Kandahar Air Field were being re-wired to meet NEC 2008. There may be a most timely and cost effective alternative to meet NEC other than a complete re-wiring.

UNCLASSIFIED

Appendix D. Management Comments to Preliminary Observations (continued)

UNCLASSIFIED

Recommendation: Determine if replacing the electrical panel, in lieu of complete re-wiring, will bring the ablution units in compliance with NEC 2008.

USFOR-A Way Ahead: There are two issues being corrected at KAF. First, the initial deliveries of ablution units from Corimec were wired using plastic electrical panels and non-GFCI receptacles over the sinks. This required the upgrade of the electrical panel and receptacles in 36 Corimec units. Corimec has been made aware of the NEC requirement and all remaining units will be corrected at the factory. Second, 32 units were received from a local supplier. The electrical wiring in these locally supplied ablution units were below acceptable standards, and all require a complete electrical wiring replacement. The local supplier is no longer being used.

CENTCOM Response: CENTCOM concurs.

13. **Observation:** The power plant at Kandahar Air Field is at maximum capacity (22 MW) and usage is expected to expand by 75% by next year.
    - Last year, peak demand was 20 MW and is expected to increase by 15 MW by next year.
    - Current location will allow for an additional generations of 12 MW.
    - According to the contractor, a plan is in place to open an additional power plant at an additional location.
    - Power shortages may require spot generation

USFOR-A Way Ahead: This is a NATO managed Prime Power provided power plant. The USFOR-A Prime Power personnel have no input or control over this plant, the operation, maintenance or required upgrade to the plant. CAPT Kuellmer is scheduling a trip there within a week and will meet with the NATO designated plant operators to obtain a plan for their way ahead.

CENTCOM Response: CENTCOM concurs. This question, although correct in it observation, has no relation to Task Force Power. Creation of COA's to solve future power requirements at KAF are for Prime Power to resolve. USFOR-A should monitor the issue for all U.S. controlled facilities in KAF.

UNCLASSIFIED

Appendix E. Management Comments
U.S. Central Command

UNITED STATES CENTRAL COMMAND
OFFICE OF THE CHIEF OF STAFF
7115 SOUTH BOUNDARY BOULEVARD
MACDILL AIR FORCE BASE, FLORIDA 33621-5101

29 June 09

TO: DEPARTMENT OF DEFENSE INSPECTOR GENERAL (DoDIG)

SUBJECT: Review of Draft DoDIG Audit Report "Assessment of Electrical Safety in Afghanistan" (D2009-D00SPO-0192.000)

1. Thank you for the opportunity to provide updates to the DoDIG report.

2. USCENTCOM has updated the responses to this report in coordination with USFOR-A. Please see the enclosure provided for further detail.

3. The Point of Contact is Colonel Mario V. Garcia, Jr., USCENTCOM Inspector General, (813)827-6660.

JAY W. HOOD
Major General, U.S. Army

Attachment:
CENTCOM Consolidated Responses

Appendix E. Management Comments
U.S. Central Command (continued)

UNCLASSIFIED

25 JUN 2009

SUBJECT: CENTCOM RESPONSE TO FINAL DRAFT, DoD IG ASSESSMENT OF ELECTRICAL SAFETY IN AFGHANISTAN (#D2009-D00SPO-192.000)

Ref: USFOR-A'S WAY AHEAD TO IG ASSESSMENT OF ELECTRICAL SAFETY IN AFGHANISTAN dtd 6 May 2009

The following are CENTCOM updated responses to the preliminary findings Way Ahead response submitted by CCJ4-E on 21 May 2009.

1. **Observation:** We observed electrical issues at Camp Brown and FOB Spin Boldak involving grounding, bonding, circuit protection, and personnel protection. Those conditions need immediate correction or will likely result in significant safety issues.

   Recommendation: Take immediate action to correct the electrical deficiencies at both locations.

   CENTCOM Way Ahead:

   Camp Brown (on Kandahar): Life Safety deficiencies were corrected on 11 May 2009. The transition from LOGCAP to Inglett and Stubbs International was abandoned due to legal implications. USFOR-A has now engaged with the component engineer to develop a Statement of Work for all electrical repairs. LOGCAP will complete the rewire/rework Statement of Work.

   Spin Boldak: All Life Safety issues have been completed. Completion of remaining electrical deficiencies is scheduled for 16 September 2009.

2. **Observation:** There is a lack of a comprehensive inventory of U.S. controlled facilities in Afghanistan.

   Recommendation: Identify and record all facilities controlled by U.S. forces in Afghanistan.

   CENTCOM Way Ahead: All existing facility data is being consolidated into an accessible data base located on the USFOR-A website.

3. **Observation:** There is a lack of a comprehensive electrical inspection plan for U.S. controlled facilities in Afghanistan.

   Recommendation: Develop a comprehensive plan to inspect, detect, and correct electrical deficiencies in facilities controlled by U.S. forces in Afghanistan.

UNCLASSIFIED

Appendix E. Management Comments
U.S. Central Command (continued)

CENTCOM Way Ahead: USFOR-A is finalizing their comprehensive plan. When finished, the plan will consist of three phases, (1) initial facility inspection and creation of an initial inspection report (currently underway), (2) analysis of the initial inspection report to determine the proper corrections, materials and funding requirements and (3) continued inspection of all maintence work and new construction.

4. Observation: There is an apparent shortage in resources to properly inspect, detect, and correct electrical deficiencies. The shortages include qualified engineers, inspectors, electricians, and funds.

   Recommendation: Complete a comprehensive electrical inspection to determine requirements. Using those results, acquire the necessary financial and personnel resources to accomplish the mission.

   CENTCOM Way Ahead: Task Force POWER continues their comprehensive inspection of existing facilities. This effort will increase with the arrival of the initial inspection team. The initial inspection team includes an organization 116 personnel comprised (eight engineers, three Master Electricians, 72 Subject Matter Experts/Electrical Inspectors and 33 support personnel). After the initial inspections are complete, these personnel will transform into the continued inspection staff totaling 42 personnel. After solicitation process is complete, the funding requirements for these services will have been determined, and the necessary funding request will be processed. Target date for their arrival is September 2009. As stated in observation three, once initial inspections are completed and their results analyzed, requirements for funding, materials and manpower will be requested.

5. Observation: Currently, the lowest level of recordable electrical accident is one that causes a non-fatal injury and causes any loss of time from work beyond the day or shift it occurred; a non-fatal occupational illness that causes loss of time from work; or disability at any time and/or property damage totaling at least $20K but less than $200K.

   Recommendation: All electrical shocks should be reported through appropriate channels.

   CENTCOM Way Ahead: Electrical shocks are now being reported through the chain of command. USFOR-A has provided direction to immediately de-energize the facility that provides the electrical shock.

6. Observation: There is no full-time cadre that is dedicated to electrical safety of U.S. controlled facilities in Afghanistan.

   Recommendation: USFOR-A needs to establish, through the Joint Manning Document process, a full-time cadre that is dedicated to electrical safety in Afghanistan.

   CENTCOM Way Ahead: USFOR-A TF POWER has defined a JMD of 10 Personnel to be added to the USFOR-A Staff for the Electrical Code Enforcement and Management. One O-6 as the CDR, four civilians (one Registered Professional Electrical Engineer, one Database Manager, one AutoCAD operator, and one clerk to manage the documentation tracking) and

Appendix E. Management Comments
U.S. Central Command (continued)

UNCLASSIFIED

five military electrical inspectors (one E-7 and four E-6 with the proper MOS/Rating and a current International Code Council certificate of Commercial Electrical Inspector E-2 credential). The submission has been delayed until completion of the 60 day assessment of the Afghanistan AOR by the newly appointed USFOR-A Commander, General McCrystal.

7. **Observation:** Multiple parties have expressed that $750,000 cap on funds for minor construction results in inefficient use of funds resulting in piece-mealing or down-scoping projects.

   Recommendation: CENTCOM, in coordination with USFOR-A, submit a legislative proposal to increase $750,000 cap on minor construction for contingency operations in Southwest Asia.

   CENTCOM Way Ahead: (02 and 03 JUN) CENTCOM Engineer ICW with OSD-C, JSJ8, and AFCENT in three different engagements briefed Congressional staffers from the HASC; SASC & SAC-MILCON; and SAC-D and HAC-D. Staffers did not support the FY10 NDAA UMMC proposal increase to $3.0M, instead recommending DOD streamline the CCA and existing UMMC request process to expedite construction execution.

8. **Observation:** There is no arbitration cell (Authority Having Jurisdiction [AHJ]) at USFOR-A with the ability to grant waivers to the NEC 2008 in Afghanistan.

   Recommendation: Establish an AHJ with the authority to review requests and grant code deviations, as necessary. The AHJ should be established at USFOR-A and in accordance with guidance in NEC 2008, Annex H.

   CENTCOM Way Ahead: CAPT Kuellmer has been established as the AHJ. He is now responsible for all electrical issue adjudication, policy development and code enforcement.

9. **Observation:** There is a lack of training for individual soldiers on the awareness of electrical hazards, the reporting process, and consequences for tampering with electrical circuits.

   Recommendation: Include training on electrical safety and incident reporting as part of pre-deployment and in-theater training. Training should also emphasize consequences for tampering with electrical circuits.

   CENTCOM Way Ahead: Creation of the training program continues. TF POWER has developed training slides to be added to the Theater RSO&I briefing program. Further training of the individual unit Training Officers will be implemented upon arrival of the USFOR-A Safety Officer and Staff after submission and fill of the JMD.

10. **Observation:** Command has identified the use of unlisted power strips and transformers.

    Recommendation: Establish a program to provide listed replacement power strips and transformers.

UNCLASSIFIED
- 3 -

Appendix E. Management Comments
U.S. Central Command (continued)

CENTCOM Way Ahead: Unlisted power strips and transformers have been replaced with UL and CE rated power strips.

11. **Observation:** A lack of qualified Contracting Officer Representatives (CORs) has been identified as a problem throughout Afghanistan. Additionally, good business practices require the separation of duties in contract execution; the lack of qualified CORs places the separation of duties principle at risk. Because of the technical requirements needed to oversee electrical contracts, the problem is particularly acute. We believe a potential remedy could be to identify individuals having technical skills and train them to be Contracting Officer Technical Representatives (COTRs) to support existing CORs.

    Recommendation: Identify and train the individuals needed to satisfy COTR requirements.

    CENTCOM Way Ahead: Anticipated arrival date in theater of USACE A&E SMEs is 16 September 2009. A&E SMEs will be trained as COTRs to collect inspection reports and develop the proper documentation for the responsible COR. DCMA has increased their theater electrical system CORs from 16 to 53.

12. **Observation:** New ablution units in Kandahar Air Field were being re-wired to meet NEC 2008. There may be a timelier and cost effective alternative to meet NEC other than a complete re-wiring.

    Recommendation: Determine if replacing the electrical panel, in lieu of complete re-wiring, will bring the ablution units in compliance with NEC 2008.

    CENTCOM Way Ahead: Changes to existing Corimec were necessary. Corimec response has been positive to correcting this issue during manufacturing prior to delivery. Correction of all discrepancies continues.

13. **Observation:** The power plant at Kandahar Air Field is at maximum capacity (22 MW) and usage is expected to expand by 75% by next year.
    - Last year, peak demand was 20 MW and is expected to increase by 15 MW by next year.
    - Current location will allow for 12 MW additional generations.
    - According to the contractor, a plan is in place to open an additional power plant at an additional location.
    - Power shortages may require spot generation

    CENTCOM Way Ahead: This is a NATO managed facility and out of the jurisdiction of USFOR-A. Should Kandahar Air Field transfer over to a US Control FOB, the issue will be fully addressed by TF POWER.

Appendix E. Management Comments
Defense Contract Management Agency

DEFENSE CONTRACT MANAGEMENT AGENCY
6350 WALKER LANE, SUITE 300
ALEXANDRIA, VIRGINIA 22310-3241

JUL 0 9 2009

IN REPLY
REFER TO   DCMA-D

MEMORANDUM FOR PRINCIPAL DEPUTY, DEPARTMENT OF DEFENSE
INSPECTOR GENERAL

SUBJECT: DODIG Draft Report, Assessment of Electrical Safety in Afghanistan, dated May 29, 2009

Reference: DODIG Project No. D2009-D000SPO-0192.000.

Attached is the Headquarters, Defense Contract Management Agency response to recommendations cited in the subject audit report.

If you have questions, please contact Mr. Harry Gilburth, DCMAI-HOCI, at (703) 428-0952 or by email: harry.gilburth@dcma.mil.

Charlie E. Williams, Jr.
Director

Appendix E. Management Comments Defense Contract Management Agency (continued)

DCMA Comments in Response to DODIG DRAFT Report, Assessment of Electrical Safety, Project N. D2009-D000SPO-0192

Recommendation 11. We recommend that USFOR-A, in coordination with DCMA, identify and train the individuals needed to satisfy COR requirements.

DCMA Comments: Agree. The following describes our Contracting Officer's Representative (COR) training process in both contingency operations.

DCMA Quality Assurance Representatives (QARs) train CORs. Prior to deployment, QARs receive training on contingency contracting operations at the DCMA Basic Contingency Operations Training (BCOT) course. During BCOT, QARs receive a COR training package that was developed by DCMA ACOs. The Theater Lead QAR is responsible for the periodic review and evaluation of the QARs' overall performance in theater, which includes effectiveness in training and utilizing CORs to support the DCMA contract oversight mission.

Personnel provided by the military services to perform COR duties are certified and qualified by their parent unit. The parent unit certifies that the nominated COR has the technical knowledge and has received the requisite COR training. The COR training/qualifications consist of administrative abilities, security clearance, no conflict of interest, available duty time, and two computer-based training classes (COR with a Mission Focus and Combat Trafficking in Persons).

In Theater, a DCMA acquisition professional provides training to the COR in a briefing and in a one-on-one format. The briefing presentation covers contract administration team roles, contract fraud, contracting terms, Procurement Integrity Act, proprietary data, unauthorized commitments, audit/inspection techniques, inspection/receiving reports, quality assurance surveillance plans (QASP), corrective action requests, contractual record keeping, COR mission, COR responsibilities, what to do, and what not to do. The one-on-one training is performed by the DCMA QAR and is accomplished during a contractor audit. The one-on-one training serves a dual purpose; provides an evaluation of the CORs capability and serves as an opportunity for the DCMA QAR to train the COR in the practical application of inspection techniques.

Appendix F. Report Distribution

Office of the Secretary of Defense

Secretary of Defense
Deputy Secretary of Defense
Chairman of the Joint Chiefs of Staff
Under Secretary of Defense for Acquisition, Technology, and Logistics
Under Secretary of Defense for Policy
Under Secretary of Defense for Personnel and Readiness
Director, Defense Procurement Policy and Acquisition

Department of the Army

Inspector General of the Army
Auditor General, Department of the Army
Commanding General, U.S. Army Corps of Engineers

Department of the Navy

Naval Inspector General
Auditor General, Department of the Navy

Department of the Air Force

Inspector General of the Air Force
Auditor General, Department of the Air Force

Combatant Commands

Commander, U.S. Central Command
       Commander, U.S. Forces-Afghanistan
           Commander, Joint Contracting Command-Iraq/Afghanistan
           Commanding General, Combined Joint Task Force-101
Commander, U.S. Joint Forces Command

Other Defense Organizations

Director, Defense Contract Audit Agency
Director, Defense Logistics Agency
Director, Defense Contract Management Agency

Congressional Committees and Subcommittees, Chairman and Ranking Minority Member

Senate Subcommittee on Defense, Committee on Appropriations
Senate Committee on Armed Services
Senate Committee on Homeland Security and Governmental Affairs
House Subcommittee on Defense, Committee on Appropriations
House Committee on Armed Services
House Committee on Oversight and Government Refor